PUFFIN BOOKS
LISTEN TO THE ANIMALS

Mr S. G. Pothan worked for the Indian Railways
for more than forty years. He published four
children story books.

Shobita Punja has worked in the field of education
for more than fifteen years, training teachers and
preparing educational material. She now writes
books on Indian Art and History.

Other Children's Books by Shobita Punja

Stories About This and That

LISTEN TO THE ANIMALS

Shobita Punja

*(As told to her by
her grandfather, S.G. Pothan)*

Illustrations by

Mario de Miranda

PUFFIN BOOKS

Penguin Books India (P) Ltd., 210, Chiranjiv Tower, 43, Nehru Place,
New Delhi 110 019, India
Penguin Books Ltd., 27 Wrights Lane, London W8 5TZ, UK
Penguin Books USA Inc., 375 Hudson Street, New York,
New York 10014, USA
Penguin Books Australia Ltd., Ringwood, Victoria, Australia
Penguin Books Canada Ltd., 10 Alcorn Avenue, Suite 300, Toronto, Ontario M4V
3B2, Canada
Penguin Books (NZ) Ltd., 182-190 Wairau Road, Auckland 10, New Zealand

First published by Penguin Books India (P) Ltd. 1994

Copyright © Shobita Punja 1994

10 9 8 7 6 5 4 3 2 1

Typeset in Palatino by Gulmohur Press, New Delhi

To my grandfather,
S. G. Pothan
and
Samiha, his great-granddaughter

Dear Friend,

The stories in this book were all written or told to me by my grandfather, S. G. Pothan. When I was a little girl he used to let me sit on his lap and he would tell me stories. They were stories about animals and people, about this and that and from here and there. The stories were about kindness and friendship, how people are and how we should behave.

You can read these stories out aloud. You can ask your parents, grandparents, elder brother or sister, to read them to you. You can make the stories into plays or puppet shows in your school. You can write stories of your own.

The beautiful pictures in this book were drawn nearly thirty years ago by Mario de Miranda. He is a famous artist of India and has illustrated many books. I am very grateful to him for allowing me to use his pictures for this book. I hope after reading these stories you will make your own drawings.

I loved my grandfather very much and I have always loved his stories. I wish that you too could hear him tell them. I have rewritten those that I love the most and offer them to you as a gift from my grandfather.

Shobita Punja

12 October 1993
New Delhi

CONTENTS

BUZZ THE BUMBLE BEE

In the garden there lived a family of bumble bees. The littlest bee was very smart with bright yellow stripes on her glossy black back. Her sisters called her 'the-little-one-who-had-a-lot-to-learn'. Her elder brothers called her Buzz, because she made a terrible buzzing sound with her wings when she was in a temper. Buzz was a brave little bee flying around the garden all on her own, inspecting everything that was curious and strange.

One day, Buzz told her mother that she wished to fly away and discover the world on her own. Her mother said, 'Listen little Buzz, your wings are not quite as strong as they should be. Stay a little longer in the hive before you start a family of bees of your own.'

Buzz would not listen and feeling very proud, buzzed loudly as she flew out of the hive.

Buzz flew past the trees in the garden and down the road when all at once she smelt a sweet scent floating past her. She followed the fragrance and the perfume grew stronger and stronger till she realized she was flying over a very big lake. Looking down at the lake Buzz saw one corner of the water's edge covered with lovely lotus flowers. The large green lotus leaves were floating on the lake like enormous saucers and the lotus flowers had raised their petalled heads high above the waves as though they were admiring the sun.

Buzz realized that the sweet perfume had come from the lotus flowers and turned herself round to make a nosedive. She landed at top speed with a heavy 'thump' on a lotus flower.

The delicate Lotus Flower trembled and shivered. Closing her petals quickly she said, 'How dare you thump and bump on my petals, you rude, little bee! You want to drink some of my sweet smelling honey, don't you? Well, I won't give you any. Some kind friend will have to teach you how to make a lotus give its honey to a bee.'

Hearing the Lotus, Buzz flew off in a huff. After flying for a while near the water's edge, Buzz saw a beetle at the foot of the tree. She went near the beetle and said, 'Tell me, Mr Beetle, what should I do to make the Lotus give some honey to me?'

The beetle was a grumpy fellow and said, 'Be off with you and your silly questions and stop buzzing around me. Can't you see I am busy making a home in the tree?'

Buzz flew into a temper and buzzed away noisily. She saw a large, silvery web hanging across the branches of the tree. Going up to the web she saw a spider spinning away. Buzz said, 'Hello, Mrs Spider-who-makes-beautiful-webs! Could you tell me what the Lotus wants from me?'

3

Mrs Spider was busy making her web. The silvery threads rolled out of her mouth as she went criss-cross from one branch to another. The web was now like a fine net ready to catch any little insect or fly that went past. Mrs Spider said, 'Perhaps the Lotus wants a fly.'

Off went little Buzz to the Lotus Flower and asked if she would like a fly in exchange for some honey. 'Have you ever heard of a lotus who eats flies?' said the angry Flower. 'Do I look like an insect to you? Flowers don't eat insects, you silly bee!' was the haughty reply.

Dismayed but not discouraged, little Buzz flew up into the sky. Flying higher and higher she looked up and saw the friendly face of a fluffy cloud.

'Hello, pretty Cloud! Do you know what the Lotus wants from me before I can drink some honey?'

'I know flowers and trees like the rain, maybe the Lotus Flower would like some raindrops,' said the helpful cloud.

Rushing back to the Lotus and jumping up and down on the petals, Buzz shouted in an excited voice : 'I know what you want! I know what you want! You want raindrops from the clouds.'

'I get all the water I need from the lake. That's why I grow on its still, calm waters. I don't need raindrops from you, little Bee.'

Very sad and a little hungry, Buzz flew away from the Lotus. Beside the lake were some lovely, old trees and through their leaves, Buzz could see ribbons of sunlight. Flying up to a sunbeam, Buzz asked, 'Golden sunbeam of speckled light! Can you tell me the secret charm that will make the Lotus share her honey with me?'

'Look out on the lake and see how the lotus flowers open their petals to the sun. See how the other bees are circling the flowers. The whole lake is bathed in sunlight. Maybe your Lotus wants some warm sunshine.'

Buzz was very pleased and went smiling all the way to the Lotus. 'I know what you want, Lotus, you need a sunbeam to keep you warm!'

'The sun is my friend and every morning when it rises in the east, I open my petals to its light and warmth. We all grow strong and big in the gentle company of my friend, the sun. I don't need warmth from you, little Bee.'

Buzz was very tired, hungry and hot. Buzz, buzz, buzz she went. As she flew away she thought, 'Who shall I ask now? Who knows the secret of the Lotus Flower? Will I always be hungry? Will I never taste the honey of the Lotus Flower that smells so sweet?'

As Buzz was flying and thinking, an old owl looked down from the tree and asked her, 'What's the matter, little Bee, you are too young to be on your own, too young to be so unhappy and too young to be so hungry.'

Buzz told the Old Wise Owl everything. She told the Owl how she had flown away from home, how she had landed with a thump on the Lotus Flower and what the Flower had said. Buzz told the Owl how she had met Mrs Spider-who-makes-beautiful-webs, how the pretty helpful Cloud had taught her that flowers like rain and how the warm friendly sunbeam had showed her that flowers love to dance in the sunlight. But no one had told Buzz what the Lotus wanted from a bee.

'Please, please help me, Wise-Owl-who-lives-in-a-tree,' said Buzz a little anxiously. 'You are a kind owl who hunts at night and knows all the secrets of the world. I know you know the way to make the Lotus give its honey to me. Please tell me, please.'

The Wise-Owl-who-lives-in-a-tree, winked once and then winked again. 'Come a little closer and I will whisper the secret in your ear.'

The leaves on the tree bent closer, the birds on the branches were quiet and everyone waited to hear the secret. No one heard it but Buzz, the bumble bee.

'Thank you very much,' said Buzz, 'I'll never forget your kindness.'

Without wasting a moment, Buzz flew toward the lake to find the Lotus. There floating on the waters was the Lotus smiling sweetly at the sun. Her white petals shone pure in the warm afternoon light.

Buzz flew down and landed on the petals as slowly as possible. She walked so lightly that the petals did not even move. Then Buzz gave the Lotus a gentle kiss. 'Sweet Lotus! Could I please share some of your delicious honey,' said Buzz.

The Lotus smiled and opened her petals so that Buzz could get to the centre where the honey was stored. 'I see you have learnt the greatest secret of all. Everyone likes a little kindness now and then. A little politeness goes a long way, my friend!'

2

HOW THE KINGFISHER
GOT ITS COLOURS

In a far-off forest there was a lovely, clear blue lake. Beside the lake were many trees, reeds and grasses. On one tree near the lake lived a bird. It was a pretty bird, the colour of rich toasted almonds, her deep reddish brown feathers glistening in the sunlight.

The bird would sit for hours on a high branch of a tree and look at the lake. When it saw the shadow of a fish swimming in the crystal clear lake, it would dive into the water and catch the fish. Sometimes it would fly into the air and hover over the waters, flapping its wings so fast that it could stay in one spot, and watch. Suspended in the air the bird could watch which way the fish were moving, then in a flash, it would dive, (beak first) into the water and catch a fish. As the bird practised all day it had become quite a good fish-catcher. Sometimes the fish got away, but that was only once in a while.

A lazy, old crow lived in the same tree as the pretty, brown bird. As summer approached the brown bird began to make her nest in the soft earth bank near the tree. High above in a fork of the tree the lazy crow made her untidy nest.

One day, when the brown bird left her nest to go fishing, the lazy old crow decided to play a trick. She changed the eggs from her nest with the ones in the brown bird's nest.

The brown bird did not notice the change till one fine morning the eggs began to crack and out came four little black crow-looking chicks. The brown bird was a little surprised (as you can imagine) but she didn't mind, because they were her chicks after all.

The next day the brown bird looked up at her neighbour's nest on top of the tree and saw that the crow had four cute brown chicks in her nest.

The brown bird knew when she saw the cute brown chicks that something was amiss. She asked the crow why her chicks did not look like crow chicks. The crow pretended to be very busy cleaning her nest and did not answer any of the brown bird's questions.

'Oh dear! I must not quarrel with the crow,' thought the brown bird. 'We live together in the same tree and if we fight with our neighbours then life will become very disagreeable.'

So the brown bird said to the lazy old crow very politely, 'Do you think we should ask the wise old-jackal-who-drinks-water-by-the-lake, why your chicks look like my chicks and my chicks look like yours?'

'Not now. I'm very busy,' said the crow.

While the brown bird was out catching fish, the crow went to meet the jackal.

'Hello, wise old jackal!' said the crow, 'I brought you a small present. I hope you like it.' The crow then dropped a piece of meat that she had found, in front of the jackal. She did this for a few days and soon the jackal and the crow became very good friends and told each other some of their secrets.

The next time the brown bird asked the crow whether the time had come to visit the jackal, she gladly agreed to go.

So the lazy old crow took the brown bird to meet the jackal.

'Dear friend! The brown bird says that I stole her eggs,' said the crow.

The brown bird bowed her head and said politely, 'Sir, when the eggs in my nest hatched, my young ones were all black and the crow's chicks look just like me, they are all brown. Perhaps, there was a mix up.'

The jackal looked very wise and sniffed the air a little and asked, 'Have you ever seen a white cow with a black calf? Have you ever seen a black dog with a black and white puppy? Have you ever seen a tall father with a short son? Have you ever seen a simple mother with a pretty daughter? A long-nosed father with a short-nosed daughter, a silly son with a clever mother?'

11

'Yes, but...' started the brown bird, a little puzzled because she knew that strange things are possible.

'Well if this is true for animals and humans then why can't it be true for birds. Go my friends, and quarrel no more,' advised the jackal.

The brown bird could do nothing and returned to her little chicks.

The brown bird had a friend, a sweet little monkey. He often visited the tree making a lot of noise to frighten the birds, laughing and shouting as he swung from the branches. She told the monkey her story. 'I think you should go to the king and complain about the lazy old crow,' said the monkey.

'Ah! But is there a king in the whole, wide world who will listen to the sad story of a little bird?' asked the brown bird.

'You are right, there are not many humans who will listen to sad stories, even fewer listen to birds, so there may not be a king who will help a poor little bird,' said the monkey.

As the brown bird and monkey were talking they heard shouts and cries from the far side of the lake.

'What's the noise all about?' asked the brown bird.

'I do not know what the noise is about,' said the monkey looking around curiously.

'Let me go and find out,' said the brown bird and flew off to the other side of the lake. The monkey went yelping and chattering as he swung from tree to tree to find out what the matter was.

At the other side of the lake the brown bird saw a huge crowd of people. There was a lovely princess in her fine clothes standing by the water and crying. Beside her were her friends and servants hurrying this way in and that way out. Everyone was shouting and screaming so loudly that the brown bird could not understand what the matter was.

Slowly, the brown bird put two and two together. The princess had come to the lake to play with her friends and while they were swimming her royal ring had slipped off her finger and no one could find it.

Word had reached the palace and the king came hurrying to the lake to comfort his daughter.

'What shall we do to find the ring?' the king asked the people who had gathered there.

One person said that the king should call a diver to find the ring. But another said, 'The diver may disturb the water and the sand below and the ring could get lost forever.'

Another person said, 'Oh king! I know a little brown bird who lives by the lake who is a very good fisher and a great diver. If you request the brown bird maybe she will find the ring without disturbing the water.'

'What a good idea! Go and call the brown bird and tell her the king wishes to speak to her!' ordered the king.

The brown bird was listening to everything and she flew out over the lake. She hovered over the place where the princess and her friends had been playing and searched for the ring. There in the water she saw the shining ring. Beak first, she dived into the water, picked up the ring and flying to the shore she dropped it beside the king. The little brown bird then flew and perched on a nearby branch of a tree. (She was a little afraid of the king, you see.)

The princess and the king and all their friends were very happy that the ring had been found.

'Well, brown bird! How can I reward a clever, helpful bird like you?' asked the king.

The brown bird told the king her sad story about the crow and the chicks.

The king was angry when he heard the story and sent for the crow.

'You are a lazy old crow,' said the king. 'You have cheated the brown bird but you cannot fool me. Be a good bird and return the chicks to their mother.'

'But they are mine,' screamed the crow.

'Return them at once! For telling a lie, I will have to punish you. From now on and forevermore, your voice will be so harsh that no one will enjoy listening to your lies.'

Turning to the brown bird the king said, 'As you have been so kind I will let you wear my royal blue colours. My shimmering pearly necklace will always hang from your chest.'

The brown bird soon got beautiful feathers of blue on her wings and on her rich brown neck was a necklace of white.

The grateful princess said, 'Thank you so much, little bird. Now you look like a king and since you are the greatest fisher I know, from now on and forevermore you will be called the White-breasted Kingfisher.'

3

THE QUICK-FOOTED, QUICK-WITTED MOUNTAIN GOAT

There was a young mountain goat who lived in a valley at the foot of the snow-capped Himalayas.

He was handsome, with his shaggy coat of long brown hair and a head crowned by a pair of strong horns. He was good-looking now but he was waiting to get his little goat beard. Then he would be perfect.

The melting snows from the mountains flowed into the streams and all the year round the waters gurgled and giggled down the valley. The valley was covered with trees, carpets of green grassy fields and scented wildflowers.

Living in such a wonderful place, the goat never went hungry. He would spend hours running up and down the valley slopes, practising his jumps and climbing up slippery rocks. He soon grew to be sturdy, a very fast runner and very quick-footed.

The goat had a large gang of friends and every morning they would meet. They would go off and play in the green fields, skipping lightly from rock to rock, feasting on the grass and flowers, and drinking at the streams. Then they would laze about and talk. Some of the goats would wander off and sleep. While the young ones would have a match and see who could climb up the hill the fastest, there were others who would find a quiet, pretty place to stand and dream.

The young, handsome mountain goat was the leader of his gang. His duty was to lead his friends to the valley each morning. He would walk on ahead and look around to make sure there were no hungry tigers around. He would sniff the air, look for danger signs and sounds of a tiger. Then he would give his friends the 'All-Clear' signal and they would happily roam the valley eating and playing all day long.

There was one person who worried all day long about the handsome mountain goat. It was his mother. She knew that there were many animals and tigers in the forests. That there were cliffs and ditches, caves and dark corners everywhere. That real danger was always present. She knew that an old tiger lived in the cave near by the valley. She knew that every two or three days the tiger got hungry and would come wandering through the valley looking for his dinner.

The mother goat told the young mountain goat to be careful. She said, 'Beware of the tiger, especially a hungry tiger. Always look to the right and then to the left. Sniff the breeze in case it is carrying the smell of the tiger. Listen for the faintest sound. The monkeys will always send a warning, the deer will bark, but a deadly silence will fill the forest when the tiger is near.'

The young mountain goat was much too young to be frightened and said, 'I know, Mother! I know the tiger would like to catch me. But he is far too old and much too lazy and I am sure I can get away even if he catches me.'

The mother goat sighed, 'You may be right but please be careful, little one. You know that a goat can never trust "someone-who-eats-goats-for-dinner." '

Every morning the young mountain goat listened to his mother's lecture. Then one day what his mother had said came true. (As it usually does.)

The young mountain goat had been playing with his friends in the valley. Everyone was tired and they were on their way home. The young goat was the last one coming down the mountain. He was making sure that no one got lost or left behind. As he reached the valley he stopped near some rocks. Suddenly, he heard a loud yawn! He looked up and there was the old tiger standing very, very near him!

It was too late to run! Too late to hide! Too late to call for help! Too late to pretend he was not there because the tiger was looking at him!

19

So the young, handsome mountain goat stood where he was with his horns lowered and ready for a fight.

The tiger looked at the funny sight and said, 'I see you are ready to fight the king of the valley. What a silly thing to do! You know that it is true that I am stronger than you. You know that I like goats for dinner. You also know that it is true that goats must never trust anyone-who-likes-to-eat-goats-for-dinner.'

The young, frightened mountain goat replied, 'I know the three truths that you have told me: that you are strong, that you like to eat goats and that goats must never trust someone-who-likes-to-eat-goats-for-dinner.'

Thinking quickly he added, 'If I tell you three truths will you let me go?'

'Okay!' said the tiger hoping that the young mountain goat would not waste too much time before dinner.

The young mountain goat said, 'Well, the first truth is that if you went back and told the other tigers of the valley that you met a mountain goat and did not kill him, they would not believe you. They would all say that you were telling a lie.'

'So true!' laughed the tiger. 'No one would believe me if I told them that I met a mountain goat and that I did not kill him for my dinner. Go on, tell me the other two truths.'

'Well!' said the young, slightly-less-frightened mountain goat, 'Suppose after our talk I go home and tell all my goat friends that I met the Great Old Tiger who lives in the valley. If I tell them that we did not fight and that you did not kill me, they would certainly never believe me.'

'So true!' said the tiger. 'I can imagine the goats would never believe you because they all know how strong and powerful I am. Go on, tell me the third truth.'

The young mountain goat thought and thought. What could he say that would save his life?

Finally, he said slowly, 'The third truth is this: since both of us are talking together and you are listening to all that I'm saying without killing me, it means that you cannot really be hungry. Is that true?'

'So true!' said the tiger, a little too quickly.

Before he realized what he had said, it was too late. The tiger remembered that he had promised to let the goat escape if he told him three truths.

So the old tiger said, 'Go away, young mountain goat, but remember that when we meet next time there will be no talking. I will catch you and you won't have the time to tell me any more truths!'

The goat ran away quickly, jumping over the rocks as fast as he could. When he was at a safe distance, he turned back and said something to the tiger.

The tiger could not clearly hear what he was saying. He thought he heard the little goat say, 'Old tiger, there is one more truth. It is this: you will never catch me a second time, I am too quick-footed and quick-witted for you!'

4

WORRIED COCKROACHES

For many years the cockroaches were worried and unhappy. They could not understand why everyone called them insects. Why everyone, the birds and animals, looked down on the whole family of cockroaches. At night when the whole world was asleep the cockroaches would come out of their hiding places and have a meeting. They grumbled that though they had wings nobody called them birds; though they had sharp claws and teeth nobody called them animals. All the cockroaches thought this was very unfair.

As worrying about such things did not help them, the older cockroaches decided it was time for some action.

They called a grand meeting of cockroaches. When they had assembled, the eldest cockroach spoke, 'My sisters and brothers! Do you know why we have gathered here today? We must find a way to stop everyone calling us insects. Most of us feel that we should be called birds or animals. If there is anyone who has a clever idea, who has something helpful to say, speak up and we shall hear him.'

Then one of the young cockroaches who was known for his cleverness got up to speak. His friends tried to make him sit down but the older one had already seen him. The eldest cockroach asked the young fellow to speak and the others to listen carefully. The young fellow began to speak.

'Sir', said he, 'I thank you for letting me speak at this great gathering. For some time, I have also thought about our problem. I have wondered how we can end our misery. I think I have found a way.'

'Hooray! Hooray! Bravo! Bravo!' shouted the gathering of cockroaches, clapping in loud applause. (Have you ever heard cockroaches clap?)

'Speak up, young fellow, let us hear what you have to say,' said someone.

Then the little fellow spoke, 'Friends, I feel that we should get one of our pretty young girls to marry an animal. If we do that, we will be related to the animals and our problem will be solved. We will have nothing to worry about.'

'How clever!' said the eldest cockroach, 'Does everyone agree to this plan?'

'Yes!' shouted everyone.

'Good, then let our elders choose a pretty cockroach girl who is willing to marry an animal. Then we will go to the mouse who lives in the next room and ask him to marry our pretty daughter,' said the eldest cockroach.

So it was agreed and the elders found a pretty cockroach who said she would be happy to marry the mouse next door. The elders decided on a time and they went off in search of the mouse.

They found the mouse sleeping in a hole in the room next door. One cockroach went up to him and tickled his nose. The mouse woke up and opened his eyes and saw a row of cockroaches standing in front of him. He said, 'Hello, cock-roaches! What can I do for you?'

'Hello, Mr Mouse. We have come to ask you something. We hope you will grant us our wish,' said the eldest of the party.

'Speak up, my friends, I am listening,' replied the puzzled mouse.

'You see, Mr Mouse,' said the eldest cockroach, 'for many years we have disliked being called insects, even though we have teeth and claws like animals and wings like birds. So we have brought our prettiest cockroach to be your wife. She would love to marry you. If you agree to marry her then we will become your relatives. Then everyone will call us animals.'

The mouse was very surprised to hear this and at the same time very sad. He said, 'I'm sorry, friends, I'm afraid I cannot help you. Though I am an animal, I am no better off than you. I live in fear all the time. I am frightened of the snake who lives in the garden. He is always waiting for me. He wants to catch and eat me. Why don't you go to the snake and ask him to marry your daughter?'

So off they went and found the hole in the garden where the snake lived. In the hole they found a magnificent cobra, the king of snakes.

A little frightened, the eldest cockroach said, 'Great snake! We have come to ask a favour of you. For many years we have been worried because we are neither animals nor birds. We went to the mouse and asked him to marry our daughter, but he asked us to come to you.'

The cobra listened very carefully and smiled. (You have seen a snake smile, haven't you?) 'I am very sorry, my friends. You see I have my own troubles. Everyone calls me a reptile. Do you know which animal frightens me most? It is that hairy fellow, the mongoose. He is very fast and very quick. He has such sharp teeth. He often fights with snakes. When he catches a snake's neck with his powerful teeth, that is the end of the snake. Go my friends, ask the mongoose to help you.'

The cockroach party went off to find the mongoose. When the mongoose heard what the cockroaches had to say he sighed, 'Friends, I thank you for this honour. Indeed your daughter is very pretty. But I am not as great as you think I am.

I am very frightened of the fox who lives in the forest. Why don't you go and ask the fox to marry your daughter?'

Though a little tired, the cockroach party went to meet the fox. He was having his afternoon nap. The cockroaches woke him up and said, 'Excuse us, Mr Fox. We have come to ask you for your help.' Then they told the fox their problem.

Hearing the story of the cockroaches the fox felt very proud. He said, 'Well, well, you want me to marry your daughter but things are not what you think they are. I am most frightened of the dog who lives in the house where you live. Whenever he sees me or smells me near his house he barks and barks and I have to run away. So I would suggest that you go and ask the dog for help.'

The cockroach party went off to find the dog who lived in the house. They were a little tired of talking and a little sad that they had not found any animal to marry their pretty daughter.

The dog listened to the cockroaches' story and said, 'Friends, it is very kind of you to ask me to marry your daughter. But I will not be a very good husband for her. You see I am the slave of the man of the house. I wear a collar made of leather around my neck. The collar is fixed on to a chain. The man holds the chain and makes me do everything he wants me to do. I have to sit when he says sit, I have to run when he says run, I can only eat when he feeds me. A dog's life is not a good life. What kind of life will your daughter have if she is married to a slave like me? Why don't you go and ask the man for his help?'

The cockroaches found the man sitting in his house. The eldest cockroach told him the whole story. How they had been worried for so long. How they had had a meeting. How they had decided to marry their daughter to an animal. They told the man how they had visited Mr Mouse and what he had said. The mouse had said that he was frightened of the snake. They told the man about their conversation with the snake and then with the mongoose, the fox and the dog.

The man listened to the long story and laughed, 'Ha-Ha-Ha! It is strange that you should come to me. Do you know the greatest worry in my life? It is you little cockroaches. I worry most about you. If I leave food out at night, you eat it. My

29

clothes have holes in them because of you. You see human beings live in fear of you, cockroaches!'

The cockroach party thanked the man and hurried away. They were all very tired but they called a meeting of all the cockroaches. When the world was asleep the cockroaches held their meeting.

The eldest cockroach explained what had happened: 'Sisters and brothers! We went looking for an animal to marry our daughter. But you see the mouse is an animal but he is frightened of the snake, who is frightened of the mongoose, who is frightened of the fox, who is afraid of the dog, who is a slave of the man. While man seems to be the cleverest of all living things and everyone is afraid of him, he lives in fear of us. Why should we be ashamed of what we are? Let us be proud of being insects, proud of being cockroaches and live happily ever after.'

'Bravo! Bravo!' shouted the crowd of cockroaches. Then everyone went home, happy with himself and happier still to be a cockroach!

5

HELP FOR THE PIGEONS

This is a story about a clever fowler, a man who used to catch birds and sell them in the market. Every day he would spread his net in the forest. He would cover the net with grass and leaves so cleverly that the birds could not see it. When the pigeons flew down from the trees in search of seeds and grass the poor birds would walk straight into the net, get entangled and caught. Every day, the fowler trapped hundreds of birds. He was very happy on the days when he caught a number of pigeons because he could sell them in the market for a good price. People would buy the pigeons to keep them in cages and to train them for pigeon fights.

The poor pigeons were very sad and worried. Every day the number of free pigeons was getting smaller and smaller. They had once been the largest flock of birds in the forest but now only a few were left.

The pigeons told their sad story to their friend, the crow, who told it to a sparrow, who told the story to a peacock, who told it to the King of Birds, Raja Hans. At once he called a meeting of all the birds in the forest. The few pigeons who were left also went to Raja Hans' meeting. When he saw the small number of pigeons left in the forest and how sad these pigeons looked he decided that something had to be done at once to help them.

The Raja said, 'Why is it that I see so many birds around me and they are all singing and chirping and you pigeons are so few in number and all of you look so unhappy?'

One pigeon replied, 'Oh, beautiful and kind Raja Hans! Long long ago we were also very happy.

Now we are worried and afraid. There is a fowler in our forest who is very clever. He traps hundreds of pigeons everyday. Now there are very few pigeons left in the forest. Raja, unless you can save us from the nets of the fowler there will be no pigeons left in the forest.'

Raja Hans turned to a large owl seated beside him and said, 'You are my chief minister and the wisest bird in the forest. You know all the secrets of the night. You have heard the sad story of the pigeons. I request you to tell us how we can help them. What can we do so that no further harm comes to them.'

Bowing low the owl replied, 'Royal Hans, I do not go out much during the day so I will not be able to help. The parakeet has very good eyes and she spends the whole day eating and playing. I am sure she will be glad to help the pigeons.'

Raja Hans turned to a flock of parakeets and asked, 'Will one of you offer your help to the pigeons?'

'Certainly, Raja sahib, I will give all the help I can,' replied one parakeet.

'Then go with these pigeons and look after them,' ordered Raja Hans.

The parakeet returned to the forest with the pigeons and told them not to worry any more.

33

Then the parakeet flew to a nearby tree and sat quietly in thought. By the evening she had worked out a plan and told the pigeons what to do.

Very early the next morning when it was still dark the fowler came to the forest. He worked quickly, quietly, and spread his net. Then he found a nice, shady bush and hid behind it and waited for the pigeons to come. But not one pigeon came to the net, not a single pigeon was seen on the ground.

All day he waited but there was no sign of pigeons. The fowler became very tired of waiting and as the sun was setting he got up to pick up his net. Just as he was about to pick it up a beautiful green parakeet with a long tail flew straight into the net. He quickly took the parakeet out of the net and was about to cut its wings so that it would not be able to fly away.

'Why are you cutting my wings and spoiling my feathers?' asked the frightened parakeet.

The fowler replied, 'If I don't cut your wings then you will fly away.'

'Oh no!' said the parakeet quickly. 'If I have all my beautiful feathers then some rich person will gladly buy me for a large sum of money. With the large sum of money you can buy food and clothes for your family.'

The fowler thought for a while. The parakeet was right. So he took the parakeet home and put her in a small iron cage for the night.

Early in the morning when he went to see the cage the parakeet said, 'Do not take me to the market in this small, iron cage covered with a dirty cloth. If you want a large sum of money, put me in a large shiny brass cage and cover it with a pretty silk cloth.'

So the fowler went to the market and bought a new cage and a pretty piece of red silk cloth. He put the parakeet in the new cage and went to the market.

All the way to the market he shouted, 'Parakeet for sale, parakeet for sale. Who will buy my pretty parakeet? Parakeet for sale, parakeet for sale.'

Many people came out of their houses and looked at the parakeet. But when the time came to buy the bird, the parakeet would not agree to go with them. She kept telling the fowler that the price was too low and that he should wait till he was offered a large sum of money.

As the fowler walked through the streets, more and more people came to buy the bird but the same thing happened. As the fowler was walking past the palace wall, the Princess opened the window from her room. She heard the fowler's cry, 'Pretty parakeet for sale! Pretty parakeet for sale!'

The Princess ran down the stairs and called out to the fowler. She lifted the pretty, red silk cloth to see the bird. There in the shiny cage sat the beautiful green parakeet. She cocked her head and said, 'Salaam, pretty Princess!'

'What a lovely bird! What a clever parakeet. She can talk so well!' exclaimed the Princess.

By this time, the Princess' elderly aunt had come down to see what her niece was doing. The excited Princess told the aunt how the parakeet had spoken. 'Please, can I have the beautiful bird?'

'Will you promise to look after it and feed it every day?' asked the aunt rather sternly.

'Yes, yes!' nodded the Princess, jumping up and down in excitement.

'How much do you want for this parakeet?' asked the old lady.

'One thousand rupees!' replied the fowler.

'That is a lot of money for a bird!' said the aunt. Then she saw the disappointed face of her niece and said, 'Here is a thousand rupees for the parakeet. Give it to us and leave this town. We do not want anyone else to have such a pretty bird.' Then she turned to her niece and said, 'Do you promise to look after this bird and feed it every day?'

'Yes, I promise!' said the Princess smiling happily.

The fowler was very happy to have made so much money. He returned to his house at the edge of the forest. When his wife saw the money she was thrilled. 'We can return to our family, to our own village and our land. We do not have to live at the edge of the forest any more.'

The family packed their belongings and started on their journey home. The pigeons were happy to see the fowler and his family leave the forest. They knew that they would now be safe, at least until another fowler come to their forest.

They were so happy that they flew high in the sky and watched the fowler and his family walking away.

Meanwhile, in the palace, the parakeet was kept in a lovely room. Every day, the Princess came to the room and fed her with delicious fruits. But though she was kind, the parakeet seemed sad. The Princess would speak to her everyday and ask, 'Why are you so sad, pretty parakeet?'

The parakeet took no notice.

The Princess thought that perhaps the bird was sad because it was locked up in the cage all day long. So she opened the cage and let her out. The parakeet was very happy. She flew up to the ceiling, around the room, sat on the table, sat on the chair and flew around the Princess. The parakeet enjoyed herself a lot and did not fly away. The Princess thought that there was no need to lock the parakeet up during the day because she would not fly away.

A few days later, the Princess thought, 'The parakeet is so happy when she is outside the cage. I will not lock her up at night any longer.'

That night when everyone was asleep, the parakeet plucked out all her old and broken feathers. She threw the feathers on the floor and flew out of the window, back to the forest.

The next morning, the Princess came into the room and found the feathers on the floor. She called out to her aunt and cried, 'Look, my poor, pretty parakeet has been killed by some cruel cat. Her beautiful feathers are lying on the floor.'

But the parakeet was safe and sound in the forest. Next day, she went to see Raja Hans and told him what had happened. He was very pleased that she had been so helpful and said, 'You are a very good and kind bird, little parakeet. You are also a very brave bird and you have saved the poor pigeons from the fowler. The fowler has left the forest and returned to his home town. I want to present you with this necklace. When the other birds see this necklace around your neck they will remember that you are a kind and brave bird who saved the pigeons.'

The parakeet wore the necklace that formed a lovely pink collar around her neck. She looked very pretty.

After a few days the parakeet thought that she should go and visit the little Princess who had been so kind to her.

She flew over the palace and saw the Princess. She was standing by the window looking sadly at the sky dreaming about her lost parakeet. The parakeet flew down and sat on the window ledge. The Princess saw her and cried with joy, 'What a lovely bird you are. You know I once had a parakeet. She was very pretty, but not as pretty as you. You have a lovely rose-ring around your neck.' The Princess then ran inside the palace and found some fruit and left it on the window ledge. Since the parakeet was not afraid of the Princess she would visit the palace every day and talk to her, eat some fruits and then fly home free and happy.

6

THE UNGRATEFUL COBRA

In a village far away there lived a young man called Rajan. His mother had died when he was young and he was brought up by his father. Rajan was a good son to his father and worked all day in the fields. Both father and son were very happy together.

In the village, there lived an old man with a young daughter called Rani. She and Rajan were good friends. They would swim in the stream together. They would play in the fields. In summer they would climb the mango trees and eat the ripe, juicy fruits. Seeing them so happy together their fathers decided that Rani and Rajan would get married when they were older.

One day, when Rajan was returning home from the fields, he saw a balloon of thick black smoke rising up in the air. He ran towards it and found that a small hut was on fire. The flames were leaping from the thatched roof and threatening the whole house. Rajan wondered what he should do. He ran first to the window to make sure that there was no one inside the hut. He saw no one. Then he looked again and saw a huge snake wriggling on the floor trying to escape from the fire. As soon as he saw the cobra Rajan broke down the door of the hut to allow the cobra escape.

Then Rajan ran to the village and called everyone to help put out the fire and save the hut. Working together they managed to save it from being completely destroyed and everyone said that Rajan had done a wonderful deed. Rani was also very proud that her friend had been so brave.

A few days later Rajan was returning home after finishing his work in the fields. He saw a huge cobra crossing his path. As soon as the cobra saw Rajan he raised his hood and was ready to strike. Rajan was very surprised and asked the cobra, 'Are you the cobra I saved the other day from the burning hut? How ungrateful you are. I saved your life and now you want to kill me?'

The cobra said, 'It is true that you saved my life the other day. But man is our most terrible enemy. Man always kills any snakes he sees, even if they mean no harm. I was only trying to protect myself by killing you. Since you have reminded me that you saved my life, let us decide what to do. Let us ask three judges to decide if I should kill you or not.'

Rajan and the snake went to look for three judges. On the way they found a cow resting under a shady tree beside a pond.

The cobra said to Rajan, 'Here are the three judges we are looking for. If you agree, we can ask the tree, the cow and the pond to be our judges.'

42

Rajan went to the tree and asked, 'Is it fair that I, who saved the cobra's life, should be bitten by him?'

The tree said to Rajan, 'I am sorry, my friend, this is the way the world is. I have been treated the same way by man. You enjoy my shade, you eat the fruit of the tree, and yet you cut my branches and give me pain. I serve you but you kill me!'

Rajan then went to the cow and explained his case to her. Then he asked, 'Cow, what do you think of this ungrateful cobra?'

The cow answered, 'Young man, I agree with the cobra because this is how man treats cows. We give you milk which you drink. With the milk you make butter, curds and all kinds of delicious sweets and ice creams. Though we feed man, he always beats us. We work hard for you in the fields but you always overwork us. We earn money for you but you do not feed us well. When we are old and no longer able to work you stop caring about us altogether.'

Rajan was very disappointed by what the tree and the cow had to say. But he knew that they were telling him the truth.

Rajan then went to the pond and asked, 'What do you think, oh pond of water?'

The pond replied, 'Without water man cannot live. You take all the water you want to drink, then you wash clothes and throw dirty things into me and spoil my clean water.'

The cobra was listening to the judges and agreed with what they were saying. He turned to Rajan and said, 'You see, man treats those who treat him well so badly. Everyone thinks that it is all right that I should bite and harm you too.'

Rajan realized that the snake was right in many ways but he did not want to die. 'I promise that I will never cut trees or beat my cows or spoil the clean water of the ponds. Please spare my life. Please, cobra, please!'

The cobra replied, 'It is impossible. From now on *you* may be kind to animals, trees and the water but what about the rest of the people. Whenever they see a snake they will kill it. They do not know what good we do to the land, how we destroy animals that destroy precious crops. Man does not know anything.'

Rajan then begged the cobra for a little more time. He knew that the cobra was an intelligent fellow. So he said, 'I want to get married to Rani, my childhood friend. Give me three years. I will marry her, have a home and family and then I will come and see you.'

The cobra was a good fellow. He could have bitten Rajan right then but he agreed to Rajan's plan.

Rajan got married to Rani and they built a house near the fields. After a year, a lovely child was born. While everyone was rejoicing and singing songs to celebrate the birth of the child, Rajan was silent and sad. Rani noticed that Rajan would often become sad and would not eat his food. Sometimes he would walk for hours all by himself in the fields and talk to the trees and the animals.

At the end of three years, Rajan grew sadder and sadder. Rani wondered what the matter was and begged him to tell her the truth. At last Rajan told her the story of how he had saved the cobra and how they had found three judges, who had explained how animals and trees and water serve man but he destroys them all. He told Rani about his promise to return to the cobra after three years. Now the time had arrived to keep his promise.

When he finished the story, Rajan thought his wife would be sad and worried. To his surprise she laughed and said, 'Do not worry! We will find a way out of this situation.'

On the last day of the third year, Rajan said goodbye to his dear father and baby. Rajan took Rani to the forest and to a dark cave where the cobra lived. He called out to the snake and said, 'Here I am with my wife. I have kept my promise and returned to you after three years.'

The cobra was delighted to see Rajan and his wife. Slowly, he crawled out of the cave and came slithering towards Rajan. He lifted his hood and was about to strike when Rani screamed, 'Not like that, King Cobra! Not like that!'

'What do you mean?' asked the cobra.

'You must come to our house with us. Then we shall set the house on fire. Then Rajan will break the door and allow you to escape. When you come out of the burning house, only then must you bite Rajan!' ordered Rani.

'What?' said the cobra, 'You want me to lie in a burning house and wait for Rajan to save me? Never! I still remember the painful burns that I had got the last time. I understand your trick now. Go away. I do not want Rajan to keep his promise. You people are too clever for me!'

So Rani and Rajan happily went back home. Every day, Rajan took his baby to the tree, near the pond. Do you know which story he told his baby?

7

THE TIGER AND THE FARMER'S WIFE

One day, a farmer went to his fields with his best bulls to plough the land. He had just begun his work when he saw a fierce tiger coming out of the forest and walking towards him.

'Salaam, my friend!' said the tiger. 'I am happy to meet you and I am happy that you have brought these two fat bulls for me to eat. This forest has become so small that there are very few animals left, so it has been a long time since I had a good meal. Now be a good man and give the two fat bulls to me quickly.'

The farmer was trembling with fear. His hands were shaking and sweat was pouring down his face. When he realized that the tiger wanted to eat his bulls and not eat him, he became a little braver and said, 'I am sorry that there is not enough for you to eat in the forest. I am sorry you have not had a good meal for such a long time. I am sorry I cannot give you my two fat bulls. I am sorry I need them to help me plough my fields so that I can feed my family.'

The tiger flew into a rage. He curled his lips and snarled, 'Enough of this "sorry" nonsense! Hurry up and give me one of your two fat bulls.'

The farmer saw that the tiger was getting angry and knew that he did not have much time. He was sure that the tiger would eat his bulls. The farmer said, 'Tiger, these bulls are old and the meat will be very tough and hard. Please wait here and I will go home and bring a fine young cow for you to eat.'

'All right!' said the tiger. 'Hurry up! I warn you, do not play any tricks with me.'

The farmer went running home. All along the way he wondered what to do. He had another worry. At home he would have to tell his wife that he had promised to give their cow to the tiger. His wife had a terrible temper and was sure to shout at him. He did not know whether he was more afraid of his wife or the tiger.

Sure enough, as soon as his wife saw him, she shouted, 'Oh, lazy bones is back again! Why have you come back to the house so early? Have you no work to do in the fields?'

'Listen to me, my dear wife,' said the farmer anxiously, 'there is no time to quarrel. I am in real danger. A huge tiger came out of the forest and asked me to give him our best bulls to eat. I have promised to give him our young cow instead. He is waiting in the fields.'

'What a clever man you are! You want to save your two bulls and give away our cow! How will the children get their milk? Without the cow we will not be able to make curds for your lunch,' said the farmer's wife.

'Let us not waste time talking about the cow.
If the bulls are killed who will plough the fields?
How will we grow the food that we eat?' replied
the husband, feeling a little braver.

'No!' said the wife. 'I will not give up my cow.
Go back, you coward and tell the tiger that you
had trouble in bringing the cow. Tell him that the
cow belongs to your wife. Tell the tiger that the
cow will only listen to her. Say that your wife
is bringing the cow for him to eat.'

The farmer returned to the field and saw the
tiger. He was in a great rage, lashing his tail this
way and that. He had been waiting a long time
to eat the cow.

'Well, well, farmer!' said the tiger. 'I don't see
the cow you had promised. So I am going to eat
your bulls. I cannot wait any longer.'

'Please, do not be in such a hurry. My wife
is on her way with the cow. Just wait a few more
minutes and she will be here. My wife will bring
the cow for you to eat, please wait a little longer.'

As soon as the farmer had left the house his
wife began to plan what she would do to save
her cow. She put on her husband's best clothes
and tied a large silk turban around her head. She

found a long, thick broomstick. Then she saddled the pony and rode to the field at great speed. When she came near the field she raised the broomstick high above her head so that it looked like a gun. She began shouting loudly, 'I am a great hunter from the west. I am told there are some huge tigers in this forest. I hope I will find a tiger soon. I want to shoot a tiger, I want to shoot a tiger!'

The tiger heard the shouts and mistaking the farmer's wife for a hunter fled into the forest.

As the tiger was running in the forest he met a jackal. 'Why are you running so fast?' asked the jackal.

'Run, jackal, run! There is a dreadful hunter on a horse and he has a gun. He wants to shoot a tiger. He may kill you as well.'

The jackal threw his head back and laughed, 'Hunter! Ha-Ha-Ha-Ha!'

'Why are you laughing?' asked the tiger.

'That is not a hunter but the farmer's wife dressed up as a hunter. She is the one who has frightened you, great tiger!' replied the jackal.

'Are you sure it is the farmer's wife?' asked the Tiger.

'Of course, I am sure. Do not be a coward. Go and kill the bulls and have a good meal and save some for me,' replied the jackal.

'I am sure you are wrong!' said the tiger. 'I saw this hunter galloping on his horse waving a long gun in his hand. He was shouting that he wanted to shoot a tiger.'

'I am not afraid!' boasted the jackal, 'I will come with you. You can see for yourself who the hunter is.'

'I do not trust you very much,' said the tiger carefully, 'you may play a trick on me. You may betray me to the hunter and run away. Then the hunter will kill me.'

'Very well,' said the jackal, 'let us tie our tails together. Then I will not be able to run away and neither will you.'

The tiger agreed and they tied their tails together. They walked slowly, as it was difficult to walk with their tails tied, back to the field. There they saw the farmer and his wife talking and laughing about the trick they had played on the tiger.

The tiger and the jackal came nearer and nearer. When the farmer saw the tiger he cried, 'Help! Help! We are going to be killed.'

'Keep quiet,' said his wife. As the animals came nearer she picked up her husband's knife and said, 'How kind of you, Mr Jackal, to bring me such a fine tiger to kill. I am very hungry. My husband and I can have this tiger for lunch.'

Hearing this, the tiger got frightened. He thought the jackal was playing a trick. So he began to run. But the tails of the tiger and the jackal were tied together. The tiger was stronger and so the jackal was dragged along. Bumpty-bumpty, bump-bump! the jackal went over the stones and thorns. Bumpty-bumpty, bump-bump! went the jackal. He howled and he cried, but it was of no use. The tiger dragged him along till they reached a safe place. The poor jackal's skin was bleeding, his back and legs were paining. He untied his tail and slowly crept away.

As he was leaving the tiger, he said very softly, 'My father was right. If you want to go forward, never trust a coward.'

8

WHY JACKALS HOWL

A very long time ago, there lived a magnificent tiger in the jungle. All the animals and birds were afraid of him especially the deer and the monkey. The animals had for a long time called the tiger the King of the Jungle.

Man had not invented the gun then nor had he learnt how to make bullets. So he too was afraid of the tiger and no one came to the jungle to shoot animals and birds. All the animals lived very happily together.

The tiger was the King, but he was very sad and lonely because he had no friends. The other animals were afraid to come close to him or to become too friendly with him. He used to watch all the animals having fun in the jungle and wished he had a friend. He watched the myna going for a ride on the elephant's back. He watched the monkeys playing in the trees and throwing fruits at the animals below. He watched the large herd of deer feeding together and running and hopping about happily in the jungle grasslands. The tiger felt very lonely and friendless.

The tiger noticed that the jackal was very cunning and clever and did not have too many friends. He thought to himself that maybe the jackal would make a good friend. The tiger called the jackal and asked him if he would like to be his friend. He took the jackal into his cave and gave him nice, juicy bones to eat.

The jackal was very happy to have become the friend of the King of the Jungle. Eating the food that the tiger left behind, the jackal became very fat and lazy.

Whenever the jackal went for a walk he would meet the monkey playing in the trees. The jackal used to boast to the monkey that he was the King's friend. The monkey did not like the boastful jackal. He would tease the jackal about his friendship with the King and make fun of him. The jackal was soon fed up with the monkey's teasing, and he decided that he would teach him a lesson.

It so happened that the tiger ate something and became very ill. He had a terrible stomach-ache and thought that he was going to die. He called the jackal and asked him to bring all the animals of the forest so that he could say his last words.

All the animals and birds came to visit their sick King. Everyone in the jungle came, except the monkey. The jackal was delighted.

Bowing very low, the jackal complained to the tiger, 'Your Majesty! Did you notice that all the animals and birds came to visit you except the monkey?'

The tiger was very angry that anyone should disobey his orders and also that the monkey did not wish to visit him when he was sick. He called the deer and said, 'Go find your friend the monkey, and bring him here at once.'

The deer dashed off to look for the monkey. and brought him to the King of the Jungle.

'Well, monkey,' said the tiger, 'why did you not come with all the others to visit me when I was sick? I have heard from the jackal that you are always making fun of him for being my friend. If that is true I will kill you, so speak up before I do.'

The frightened monkey was sitting on a tree and trembling because he knew that the tiger could catch and kill him. The whole branch was shaking and the leaves were rustling as the poor monkey shook with fear. The monkey gathered all his courage and said very sadly, 'Your Majesty, when I heard that you were ill, I was very sad. I thought you were going to die. Then I remembered that my father had a medicine that could cure any pain. The medicine is made of herbs and sweet smelling plants. I was determined to find it for you.'

The tiger was very pleased that the monkey was trying to help him. He said, 'Well, did you find the herbs and make the medicine? My stomach has not stopped aching and I am going to die.'

'Yes, Your Majesty!' replied the monkey respectfully. Then from the corner of his eye, he saw the jackal smiling cunningly at him. The monkey said, 'But, Your Majesty! I have a problem. I need a special skin to wrap the herbal medicine in for a day. I have not yet found this special skin. Without this skin the medicine will not work.'

'What is this special skin?' roared the tiger. 'Tell me at once and I shall order that it be found for you.'

'Your Majesty, I need the skin of a jackal to wrap up my herbs.'

'Tear off a piece of the skin of the jackal!' roared the tiger to the other animals.

The jackal was very ashamed of what he had done to the monkey. But it was too late. They tore off a piece of skin from the jackal and wrapped the herbs in it. The tiger took the medicine and was cured.

There was great rejoicing in the jungle when the tiger was cured.

All the animals were very happy. The jackal slipped away into the jungle. He never spoke to the monkey again, never showed off in front of anyone. He only came out at night to cry and howl with shame at his mistake. Sometimes even the moon looks down and laughs at the jackal and this makes the poor jackal howl even more.

9

THE PROUD ELEPHANT

The tiger was the King of the Jungle. He was as kind as he could possibly be to all the animals. He was very just and fair. He had no favourites and treated everyone the same. As the years went by the Tiger became old and feeble. When he was about to die he called all the animals and asked them to choose a new king.

The animals looked around and found that the elephant was the largest animal in the jungle, so they chose him as their new king.

This elephant was very proud, and became very big headed when he was chosen to be the king. He was not a very nice elephant. After the tiger died, he began to worry all the smaller animals. Everyone lived in fear of the King Elephant. He would go to the lake and bathe everyday. He would splash in the water, roll around and make the water very muddy. No one could drink from the lake for a long time after the King Elephant had had his bath. The elephant loved to rush around the jungle. When he did this he crushed many small trees and plants. Even the happy monkeys were unhappy with the King Elephant. Monkeys love to eat tender shoots and leaves of trees, they love bananas and fruit but so did the elephant. Everyone was afraid of the King Elephant and there was no more happiness in the jungle.

It was nearing the end of summer, the ground was dusty and it was very hot. The monsoon rains would soon arrive. The little black ants were working very hard from morning till night. Some of them carried food to store for a rainy day. The other ants worked hard to build the anthill over their house to keep the rain water from rushing in and drowning them.

One very hot day, the proud elephant was returning from his long swim in the lake. He saw hundreds of ants working very hard. He was very angry and said, 'When the King walks by you must all stop work and wish me.'

'We are very sorry, King Elephant, but we have so much work to do we did not see you walking by.'

'What work do you have to do, you useless, tiny creatures? With one kick I can kill all of you and break your house down.'

The ants got very worried for they could smell the monsoons coming. They knew that even if the elephant spared them the rains would not.

The bravest of them said, 'Your Majesty, we are small, even tiny, if you say so. But we are not useless. We eat all the dead insects and creatures that no one wants and keep the jungle clean. We work all day and worry no one.'

The elephant was not pleased with the ant, who was so tiny that he could hardly see it. He lifted his trunk and trumpeted and said, 'You are so small, that you must be useless. Being so tiny you must be good for nothing. I am big and I am the best.'

The brave ant said, 'Your Majesty, just because we are small it does not mean we are useless. We are tiny but we can do anything you can do.'

'Can you run a race with me?' asked the King Elephant looking down at the ant, who was not even as big as the hair on his little toe.

'Yes, we can beat you in a running race, King Elephant! If you would like to have a race with us, let us, but please do not call us useless,' cried the ant.

The elephant trumpeted madly. He was very angry and said, 'You cheeky little fellow! How can you win a race with me? It will take you forever to run a few steps of mine. I will teach you a lesson for being so small and so cheeky. We will have a race and when I win I will break your house down.'

'Your Majesty, let us begin the race.'

'Ready... Steady... Go.'

Off went the King Elephant and the ant. After a while, the proud elephant looked down and to his surprise he saw the ant. 'Hello! You are still running with me?' he said.

The King Elephant ran a little further and when he looked down he saw that the ant was still running near him. The elephant ran on and on, up the hill, down the hill, over the rocks, down the paths. Every time he looked down the ant was still following him. The King Elephant was getting very tired and hot and began to puff and pant. But every time he thought he would stop and finish the race he saw the ant near him.

The King Elephant was too proud to lose the race, so he kept on running. He raised his trunk and made loud, angry sounds as he charged through the forest. The other animals came out to see what the noise was about. When the animals found out that the King Elephant was running a race with the ant they all began to cheer, 'Come on ant, keep on the race! Come on tiny ant!'

The cheering and shouting made the King Elephant quite mad. He looked up and saw all the animals laughing at him. The elephant did not want to be beaten in the race especially by an ant so he kept on running. His legs began to ache and soon he could run no more. Sweat was pouring down his face so he could not see and his head was bursting with anger. Blind with rage, the King Elephant fell into a huge ditch and tumbled right to the bottom.

He looked up and saw all the animals looking down at him from the mouth of the ditch. The ant stood up and waved at the elephant and cried, 'I won the race! I won the race! The King Elephant was beaten by a tiny ant. Sorry but this tiny ant cannot help you out of this ditch.' Laughing loudly the ant went away.

The elephant struggled to get out while all the animals stood watching. They did not help him till he promised to be kind and fair. The elephant promised to bathe only in the evenings so that the other animals could have clean water to drink in the mornings. He promised to leave a few fruits and fresh shoots for the monkeys.

When the elephant got out, he plucked a nice banana from the tree and ate it. He plucked one banana for the monkey and asked him, 'I want to ask you something, monkey dear! Tell me, how did that tiny ant win the race?'

The monkey took the banana and said, 'You see, proud elephant, you did not take any notice of the ant. Every time you looked down you saw an ant running but it was not the same ant. There were hundreds of ants, running up and down doing their work!'

10

THE JACKAL'S CLEVER WIFE

Deep in the forest, there lived a jackal with his wife and three children. The jackal was lazy. He was too lazy to find food, too lazy to run, or to clean himself, or to even think! He just liked to sleep all day and roll in the mud. The jackal's wife was very sad because she had to find food for herself, her three children and her lazy husband. Every evening she would say to him, 'Why don't you go and find some food for us to eat.'

'Go away, I am resting,' would be the jackal's reply.

Once in a while the jackal would get up. He was too lazy to go to the forest in search of food so he would go to the nearby village. In the light of the moon he would search for something to eat from the scraps that he found in the village.

Sometimes he was lucky. He would find a nice juicy bone and bring it home to his wife and boast, 'Those stupid village dogs could not find this bone. I am so clever.'

The jackal's wife would reply, 'Yes, you are very clever!'

The jackal would say, 'If I was not so clever you and our three children would have nothing to eat and would starve to death.'

The jackal's wife and the three children were tired of hearing the jackal boast all the time. They kept quiet because there was nothing else they could do.

As time went on, the jackal became fatter and fatter because his wife continued to feed him and he got no exercise. Even when he slept, he would say in his dreams, 'I am the cleverest animal. There is no one as clever as I.'

After some time he began to say, 'As there is no one as clever as I, I think I am better than that foolish old lion!'

One night, he was out with his wife to take a stroll and smell the lovely summer evening. They were in a dark corner of the forest when who should they meet but the old lion.

The boastful jackal was very frightened when he saw the lion. He could not do or say anything. The lion came up close and said, 'I see you cannot escape. Well, jackal, you have grown fat and big. I think the time has come for me to kill you.'

The jackal was too frightened to say anything. So his poor wife had to think of something to say. She said quietly, 'Your Majesty, you are kind and fair. If you want to, please eat us. But before you came here, my husband and I were having a quarrel. Perhaps you can help us before you eat us?'

'What was the quarrel about?' asked the lion.

'We were quarrelling about our three children,' said the jackal's wife.

'You have three children?' asked the lion, and his mouth began to water.

'Yes, we have three children. We can take you to our hole where we live. You can see the three little children and then we can tell you about our quarrel,' said the jackal's wife.

'Come on then. Do not waste any time!' ordered the lion. The three of them went together to the jackal's hole. The lion thought that he was very lucky. He had a good chance of eating the jackal, his wife and the three little jackal cubs.

The jackal was very frightened. He thought his wife was very stupid to have invited the lion to their hole. His wife was thinking of nothing but how to save her little children and her husband. When they came to the hole where the jackals lived, they stopped outside. The jackal's wife said to her husband, 'Go inside quickly and bring our three children out, please.'

The jackal thought, 'My wife is not only stupid, she is mad!'

When the jackal went inside the hole, the jackal's wife and the lion waited outside. After a little while, the jackal's wife said to the lion, 'I am sorry that my husband is keeping Your Majesty waiting. I think I had better go inside and find out what the matter is. Maybe the children are giving him some trouble. If you will please wait for a minute, I will go and bring them all out.'

'All right!' said the lion. 'But do not keep me waiting too long. I am in a hurry and want to eat a jackal soon.'

The jackal's wife said nothing but went quickly into the hole. When she was safely inside she shouted to the lion, 'Your Majesty, we have settled our quarrel. You do not have to wait any longer. Thank you for trying to help us!'

The lion was very angry. He knew that he had been fooled by the jackal's wife. He also knew that he could do nothing. He was too big to fit into the jackal's small hole. So he went away in a bad temper.

After a while, when they were sure the lion had gone, the jackal family came out. The jackal cubs began to play with each other. The jackal's wife sat cleaning her paws while the jackal lay on the ground and made himself comfortable. In a while the jackal said, 'I am the cleverest animal in the world!'

The jackal's wife was very surprised; even the cubs were surprised. 'What? You were so frightened when you saw the lion you could not figure out what to do. How can you boast that you are the cleverest animal?'

The jackal smiled and said, 'My dear, I *am* the cleverest animal. I am clever because I married a clever wife.'

The jackal's wife was very happy. At last her husband thought that she was clever!